Good Science—*That's Easy to Teach*

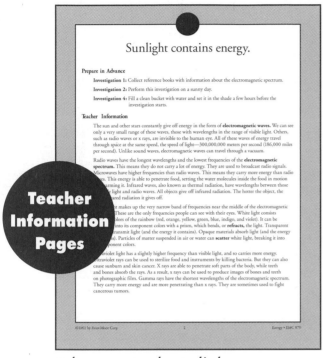

- the concept to be studied
- items to obtain or prepare in advance
- background information

Teacher Information Pages

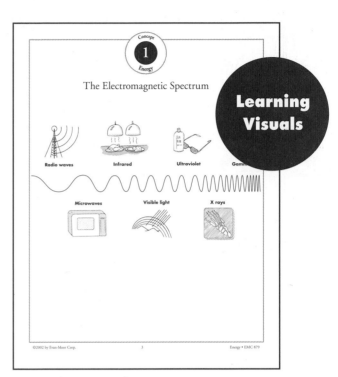

- reproduce or make into transparency

Learning Visuals

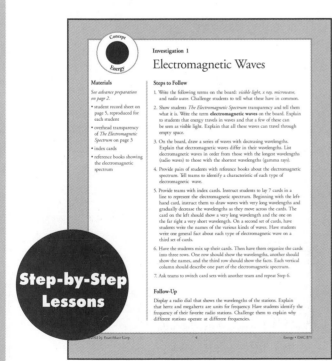

Step-by-Step Lessons

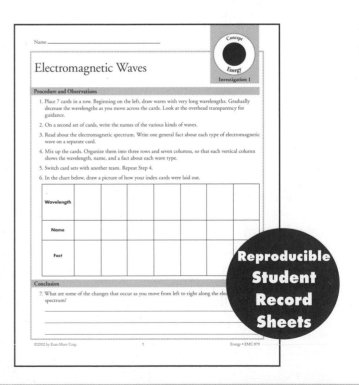

Reproducible Student Record Sheets

Sunlight contains energy.

Prepare in Advance

Investigation 1: Collect reference books with information about the electromagnetic spectrum.

Investigation 2: Perform this investigation on a sunny day.

Investigation 4: Fill a clean bucket with water and set it in the shade a few hours before the investigation starts.

Teacher Information

The sun and other stars constantly give off energy in the form of **electromagnetic waves.** We can see only a very small range of these waves, those with wavelengths in the range of visible light. Others, such as radio waves or X rays, are invisible to the human eye. All of these waves of energy travel through space at the same speed, the speed of light—300,000,000 meters per second (186,000 miles per second). Unlike sound waves, electromagnetic waves can travel through a vacuum.

Radio waves have the longest wavelengths and the lowest frequencies of the **electromagnetic spectrum.** This means they do not carry a lot of energy. They are used to broadcast radio signals. Microwaves have higher frequencies than radio waves. This means they carry more energy than radio waves. This energy is able to penetrate food, setting the water molecules inside the food in motion and warming it. Infrared waves, also known as thermal radiation, have wavelengths between those of visible light and radio waves. All objects give off infrared radiation. The hotter the object, the more infrared radiation it gives off.

Visible light makes up the very narrow band of frequencies near the middle of the electromagnetic spectrum. These are the only frequencies people can see with their eyes. White light consists of all the colors of the rainbow (red, orange, yellow, green, blue, indigo, and violet). It can be broken up into its component colors with a prism, which bends, or **refracts,** the light. Transparent materials transmit light (and the energy it contains). Opaque materials absorb light (and the energy it contains). Particles of matter suspended in air or water can **scatter** white light, breaking it into its component colors.

Ultraviolet light has a slightly higher frequency than visible light, and so carries more energy. Ultraviolet rays can be used to sterilize food and instruments by killing bacteria. But they can also cause sunburn and skin cancer. X rays are able to penetrate soft parts of the body, while teeth and bones absorb the rays. As a result, X rays can be used to produce images of bones and teeth on photographic film. Gamma rays have the shortest wavelengths of the electromagnetic spectrum. They carry more energy and are more penetrating than X rays. They are sometimes used to fight cancerous tumors.

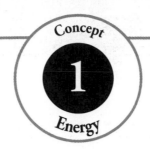

The Electromagnetic Spectrum

Radio waves

Infrared

Ultraviolet

Gamma rays

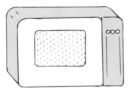

Microwaves

Visible light

X rays

Energy • EMC 879

Concept **1** Energy

Investigation 1

Electromagnetic Waves

Materials

See advance preparation on page 2.

• student record sheet on page 5, reproduced for each student

• overhead transparency of *The Electromagnetic Spectrum* on page 3

• index cards

• reference books showing the electromagnetic spectrum

Steps to Follow

1. Write the following terms on the board: *visible light, X ray, microwave,* and *radio wave.* Challenge students to tell what these have in common.

2. Show students *The Electromagnetic Spectrum* transparency and tell them what it is. Write the term **electromagnetic waves** on the board. Explain to students that energy travels in waves and that a few of these can be seen as visible light. Explain that all these waves can travel through empty space.

3. On the board, draw a series of waves with decreasing wavelengths. Explain that electromagnetic waves differ in their wavelengths. List electromagnetic waves in order from those with the longest wavelengths (radio waves) to those with the shortest wavelengths (gamma rays).

4. Provide pairs of students with reference books about the electromagnetic spectrum. Tell teams to identify a characteristic of each type of electromagnetic wave.

5. Provide teams with index cards. Instruct students to lay 7 cards in a line to represent the electromagnetic spectrum. Beginning with the left-hand card, instruct them to draw waves with very long wavelengths and gradually decrease the wavelengths as they move across the cards. The card on the left should show a very long wavelength and the one on the far right a very short wavelength. On a second set of cards, have students write the names of the various kinds of waves. Have students write one general fact about each type of electromagnetic wave on a third set of cards.

6. Have the students mix up their cards. Then have them organize the cards into three rows. One row should show the wavelengths, another should show the names, and the third row should show the facts. Each vertical column should describe one part of the electromagnetic spectrum.

7. Ask teams to switch card sets with another team and repeat Step 6.

Follow-Up

Display a radio dial that shows the wavelengths of the stations. Explain that hertz and megahertz are units for frequency. Have students identify the frequency of their favorite radio stations. Challenge them to explain why different stations operate at different frequencies.

Name ____________________________________

Concept 1 Energy

Investigation 1

Electromagnetic Waves

Procedure and Observations

1. Place 7 cards in a row. Beginning on the left, draw waves with very long wavelengths. Gradually decrease the wavelengths as you move across the cards. Look at the overhead transparency for guidance.
2. On a second set of cards, write the names of the various kinds of waves.
3. Read about the electromagnetic spectrum. Write one general fact about each type of electromagnetic wave on a separate card.
4. Mix up the cards. Organize them into three rows and seven columns, so that each vertical column shows the wavelength, name, and a fact about each wave type.
5. Switch card sets with another team. Repeat Step 4.
6. In the chart below, draw a picture of how your index cards were laid out.

Wavelength							
Name							
Fact							

Conclusion

7. What are some of the changes that occur as you move from left to right along the electromagnetic spectrum?

____

____

____

Concept 1 Energy

Investigation 2

Colors of Light

Materials

See advance preparation on page 2.

- student record sheet on page 7, reproduced for each student
- colored markers or crayons
- prisms
- sheets of white paper

Steps to Follow

1. Ask students what part of the electromagnetic spectrum they see when they look out the window (visible light).

2. Write the word **refraction** on the board. Explain that when light moves from one kind of matter to another, its speed changes and the light bends. Light of different wavelengths is bent different amounts.

3. Hold up a prism. Ask a volunteer to tell what it is. Explain that when light passes through a wedge-shaped piece of glass, it is bent, or refracted.

4. Ask students to predict what they will see when visible light is refracted by the prism. Have them record their predictions.

5. Darken the room. Pull the window shades almost all the way down. Leave only a narrow slit for light to pass through.

6. Have students place a sheet of white paper on a desk next to a window. Instruct them to hold a prism in the sunlight. Have the students move the prism back and forth until the various colors appear on the white paper.

7. Tell students that this rainbow of colors is called the **spectrum.** Challenge students to explain why all these colors appeared. (Each color is a different wavelength of visible light and each was refracted a different amount. Red light bends the least because it has the longest wavelength. Violet light has the shortest wavelength, so it bends the most.)

8. Have students use colored markers or crayons to record the order of the colors on their record sheets.

9. Challenge students to identify which wavelengths of light were refracted or bent the most when they passed through the prism.

Follow-Up

Provide students with bubble mixture and plastic wands. Have them blow bubbles in bright sunlight. Ask them to note what they see on the surface of the bubbles. (They should see many different colors. Light is reflected from the different soapy film layers. The thickest film layers reflect red light and the thinnest layers reflect violet light.)

Name ______________________________

Colors of Light

Concept 1 Energy

Investigation 2

Prediction

1. What do you think you'll see when visible light is refracted by a prism?

____

Procedure and Observations

2. Place a sheet of white paper on a table beside a window.
3. Hold a prism in sunlight. Watch the light pass through the prism and onto the paper. Draw what you see on the paper.

Conclusions

4. What happens to visible light when it passes through a prism? Why?

____

____

5. Which wavelengths of light were refracted the most as they passed through the prism? How can you tell?

____

____

6. List all the colors of light in order from longest to shortest wavelength.

____

____

____

Investigation 3

How Light Behaves

Materials

- student record sheet on page 9, reproduced for each student
- clear jars
- flashlights
- milk
- water

Steps to Follow

1. Divide students into small groups. Distribute a flashlight and a jar of clear water to each group.

2. Ask students to predict what they will see when they shine the light through the jar of clear water. Have them record their predictions on their record sheets.

3. Darken the room. Have students shine their flashlights through the jars of clear water. Instruct students to look at the water in the jar from all angles before recording their observations on their record sheets.

4. Hold a class discussion and introduce the term **transmission.** Explain that light is transmitted through clear objects. That is, it passes right through them.

5. Now challenge students to predict what will happen when they shine the light again through a beaker that contains milky water. Have them record their predictions on their record sheets.

6. Have students place a few drops of milk into their jars of water and stir.

7. Darken the room again. Instruct students to shine their flashlights through the water as before. Have students observe the jar from all angles before recording their observations on their record sheets.

8. Discuss student observations. (Depending on where they were standing, students may have seen orange or blue light in the water.)

9. Introduce the term **scattering.** Explain that as light passes through air or water with tiny particles of matter in it (like milk or dust), the particles absorb the energy of the light and then release it, scattering it in all directions. This is why students saw different colors in the light depending on which angle they were viewing it from.

Follow-Up

Have students research the explanation for why the sky is blue. (Blue light scatters the most of any color.)

How Light Behaves

Predictions

1. What do you think you will see when you shine a light through a clear jar of water?

_____

2. What do you think you will see when you shine a light through milky water?

_____

Procedure and Observations

3. Have one member of your group shine a flashlight through the jar of clear water.

4. Look at the water. What color or colors of light do you see?

_____

_____

5. Now mix a few drops of milk into the water.

6. Have one group member shine the flashlight through the jar of milky water. Look directly at the beam of light as it passes through the water. What color or colors of light do you see?

_____

7. Look at the jar from the side. What color or colors of light do you see?

_____

Conclusions

8. What happens to light when it passes through clear water and glass?

_____

_____

9. What happens when light passes through water with particles of matter in it?

_____

_____

Investigation 4

Sunlight and Heat Energy

Materials

See advance preparation on page 2.

- student record sheet on page 11, reproduced for each student
- bucket of water
- black construction paper
- white construction paper
- clear tape
- clear jars with lids
- scissors
- thermometers
- watch or clock with a second hand

Steps to Follow

1. Ask students whether they prefer to be in the sun or in the shade on a hot summer day. Invite them to explain their answers. Tell students they will set up an experiment to find out more about the energy in sunlight.

2. Divide students into pairs. Give each team two identical jars with lids. Invite each team to fill their jars to the same level with water from the bucket.

3. Provide each team with a thermometer. Have them stir the water in each jar and then measure and record the temperature of the water in each jar. (Temperatures should be equal.) They should also record the time.

4. After students have replaced the jar lids, give each team a piece of black construction paper and a piece of white construction paper. Have students cover one jar and lid with black construction paper and the other jar and lid with white construction paper. Tell them to hold the paper in place with tape. Instruct them to place their jars in bright sunlight.

5. Ask students to predict what will happen to the temperature of the water in each jar. Have them write their predictions on their record sheets.

6. Every 5 to 10 minutes, have students stir the water in each jar and then measure and record the temperature. Remind them to record the time as well. If necessary, have students move the jars so that they stay in the sunlight.

7. Continue taking readings until water temperatures start to plateau.

8. Once the last temperature reading has been taken, discuss what happened to the water temperature in the jars. Challenge students to explain what heated the water (energy in sunlight). Invite students to explain the differences in temperature between jars. (Black paper absorbs more light energy than white paper.)

Follow-Up

Instruct students to create a bar graph of their temperature data. Assist them as needed.

Name _____

Sunlight and Heat Energy

Prediction

1. What do you think will happen to the temperature of the black jar of water when it is placed in the sunlight? To the white jar?

   _____

   _____

   _____

Procedure and Observations

2. Fill each jar with the same amount of water from the bucket. Stir.

3. Measure and record the temperature of the water in each jar under "Start" on the chart below. Also record the time.

4. Cover one jar with black paper and one jar with white paper.

5. Place the jars in sunlight. Measure and record the water temperature and time every 5 to 10 minutes.

Time

	Start							
Temperature in Black Jar (°C)								
Temperature in White Jar (°C)								

Conclusions

6. What happened to the water temperature in the jars? Which jar got the hottest?

   _____

7. How can you explain your results?

   _____

   _____

2

Sound is produced by vibrations.

Teacher Information

A **wave** is a disturbance that travels though matter or space. Waves require energy for their production, and they carry that energy from one place to another.

Sound waves are created when matter vibrates. The vibrations then travel outward from the source. Sound waves can only travel through matter. They cannot exist in a vacuum. Most of the sounds we hear travel through air. However, sound waves can travel through solids, liquids, and gases. Sound waves travel faster through solids than through gases, like air.

Sound waves are caused by vibrations. For example, when a rubber band is plucked, it vibrates back and forth. As it moves upward, it pushes against the air molecules above it, causing them to be crowded together. This action results in areas where air molecules are closer together. These areas are known as **compressions.** The compressions are bordered by areas where air molecules are farther apart. These areas are known as **rarefactions.** As the rubber band continues to vibrate, more of these rarefactions and compressions form and move outward from the rubber band. They in turn affect the air molecules that they bump into, resulting in a series of alternating rarefactions and compressions moving away from the source. These rarefactions and compressions form **longitudinal waves.** When they strike our ears, we perceive them as sounds.

Wavelength is the distance from a point on one wave to the equivalent point on the next wave. For example, it is the distance from the beginning of one rarefaction to the beginning of the next rarefaction. Sound waves produced by slow vibrations have long wavelengths. Waves originating from fast vibrations have short wavelengths.

Frequency is the number of waves produced per second. Frequency and wavelength are inversely related: the higher the frequency, the shorter the wavelength; the lower the frequency, the longer the wavelength. The pitch of a sound is determined by its frequency. The greater the number of vibrations per second, the higher the frequency and the higher the pitch of the sound.

The **amplitude** of a wave is a measure of how high and low it oscillates from its resting position. Amplitude is therefore a measure of the amount of energy a wave contains. Waves with more energy have a higher amplitude than those with less energy, and they sound louder to our ears.

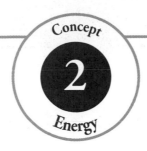
Sound Waves

Longitudinal (Sound) Wave

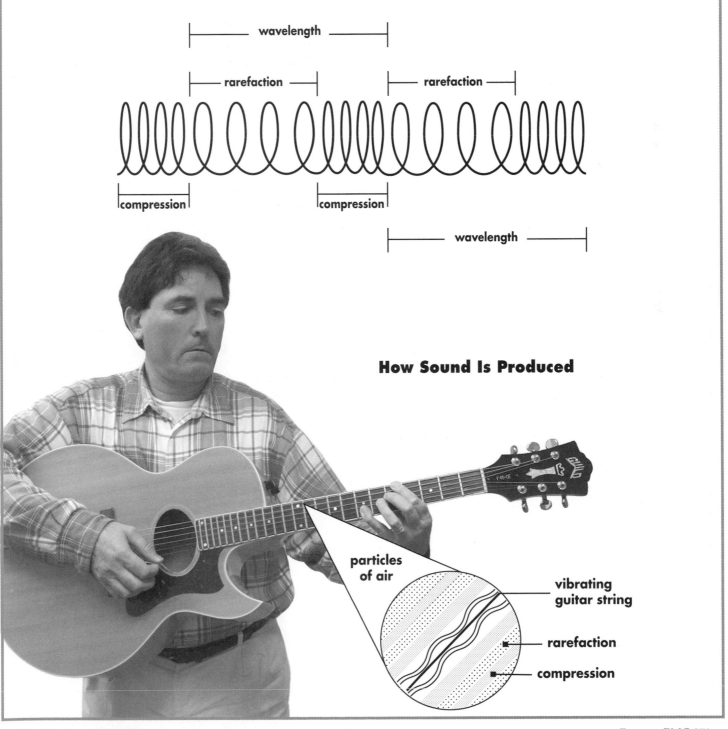

wavelength

rarefaction   rarefaction

compression   compression

wavelength

How Sound Is Produced

particles
of air

vibrating
guitar string

rarefaction

compression

Investigation 1

Making Sounds

Materials

- student record sheet on page 15, reproduced for each student
- bowls
- combs
- craft sticks
- tuning forks
- water
- waxed paper

Steps to Follow

1. Invite students to hum loudly. Have them hold a hand on their vocal cords while they are humming. Ask them what they notice and how they think sounds are produced. Have them write their ideas on their record sheets.

2. Divide students into pairs. Give each team a tuning fork. Allow them time to investigate how to create a sound with it. Make sure students discover that they can strike one prong of the tuning fork against another object and then hold it close to their ear to hear a sound. Have them strike the tuning fork and then press it lightly to their palm. Have them observe and record what they feel.

3. Distribute the other materials to students. Instruct students to tap the tuning fork and then bring it slowly toward a bowl of water until it just touches the water. Have them record what they observe.

4. Have each team fold a piece of waxed paper and place it over a comb. Then tell them to purse their lips slightly so that a thin stream of air comes out as they hum. Instruct them to hold the comb with the waxed paper just in front of their lips (but not touching them) as they continue to hum. Tell students to record what they hear and feel.

5. Have students extend a craft stick over the edge of a table and hold it in place. Instruct them to gently snap the end that sticks out and record what they hear and see.

6. Provide time for students to study their results. Discuss what is similar about all their observations. (They all involved vibrating objects.) Ask them how they think the sounds were created. (Vibrating objects create sound.)

7. Discuss what is needed to cause objects to vibrate and produce a sound. (Energy is needed to create vibrations and sound.) Lead students to see that sound is a form of energy.

Follow-Up

Provide the opportunity for students to see other kinds of vibrations that produce sounds. Show them how a plucked guitar string or rubber band vibrates. Have them feel the top of a toy drum after hitting it and blow on a strip of paper to make it vibrate and produce a sound.

Making Sounds

Prediction

1. How do you think sounds are produced?

_____

Procedure and Observations

2. Strike the tuning fork against a hard object. Press it lightly to your palm. What do you feel?

_____

3. Strike the tuning fork. Bring it slowly toward a bowl of water until it just touches the water. What do you see?

_____

4. Fold a piece of waxed paper and place it over a comb. Hum and blow gently onto the paper. What do you hear and feel?

_____

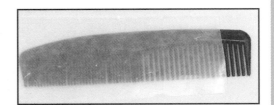

5. Extend a craft stick over the edge of a table. Hold it in place. Gently snap the end that sticks out. What do you hear and see?

_____

Conclusions

6. What did all your observations have in common?

_____

_____

7. How do you think sounds are produced? What do you need to supply?

_____

_____

Investigation 2

Sound Waves

Materials

- student record sheet on page 17, reproduced for each student
- overhead transparency of *Sound Waves* on page 13
- masking tape
- Slinkys®

Steps to Follow

1. Clap your hands. Ask students how the sound reached their ears. Review with students what they observed when they touched a tuning fork to water. Explain that sound, like other kinds of energy, travels in waves.

2. Tell students that they will investigate sound waves and how they behave. Point out that many forms of energy, including light, move in waves.

3. Divide students into pairs. Give each team a Slinky®. (Warn students that they can be easily damaged and that they should be careful not to permanently bend the coils.)

4. Tell students to stretch the Slinky® out on the floor to a length of 10 to 13 feet (3 to 4 m). Instruct students to hold it firmly at each end so that it does not move. Then have one student suddenly push one end of the Slinky® in and then quickly back out. Tell students to observe what happens. (A wave of compressed coils will move along the length of the Slinky®.) Explain that this motion is known as a pulse.

5. Have students again hold the Slinky® still. Instruct one student to push the spring in and out several times to create several pulses. Invite a third student to watch carefully from the side. Ask this student to describe the position of the coils in a pulse and in the space between pulses. Explain that a region where the coils are close together is called a **compression** and a region where the coils are far apart is called a **rarefaction.**

6. Provide masking tape, and instruct students to attach a small piece to one coil of the spring and then send a pulse along the Slinky®. Ask them to record what they observe on their record sheets.

7. Show students the *Sound Waves* transparency. (Show only the top portion of the diagram.) Tell students that the distance from one point in a pulse to the same point in the next pulse marks the wavelength of the wave. The number of pulses or waves that move by a certain point each second describes the frequency of the wave. Challenge students to increase and decrease the wavelength in the Slinky®. Have them observe what happens to the frequency. (As wavelength decreases, frequency increases.)

8. Explain that the wavelength and frequency of a sound wave determines how it sounds to our ears. Students will learn more about this in the next few investigations.

Name ______________________________

Concept 2 Energy

Investigation 2

Sound Waves

Procedure and Observations

1. With a partner, stretch out a Slinky® on the ground. Quickly push in and pull out one end. What happens to the Slinky® coils?

____

____

2. Create several pulses. Draw the appearance of the Slinky® in the space below.

3. Compare the motion of the coils to the motion of a pulse in the Slinky®. What do you notice?

____

____

4. Fasten a piece of tape to one coil of the Slinky®. Send pulses along it. Watch how the tape moves. Record your observations.

____

____

5. Try to increase and decrease the wavelength of the pulses. How does this affect the frequency?

____

____

Conclusion

6. Based on your observations of the Slinky®, how would you describe the relationship between wavelength and frequency?

____

____

____

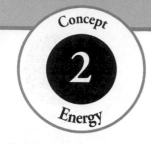

Investigation 3

Sound and Solids

Materials

- student record sheet on page 19, reproduced for each student
- bell
- spoons
- 1-yd (1 m) lengths of string

Steps to Follow

1. Ring a bell or make another loud sound. Ask students to explain how the sound reached their ears. Make sure they understand that the bell created vibrations, which in turn created compressions and rarefactions in the air, known as sound waves. Remind them that the sound waves moved through the air to reach their ears.

2. Tell students they will now investigate the movement of sound through materials other than air. Divide students into pairs. Give each team a spoon and approximately 1 yard (1 m) of string. Instruct them to use a double knot to tie the middle of the string to the spoon.

3. Instruct one student to hold both ends of the string and let the spoon hang down. Tell the other student to tap the spoon with a pen or pencil. Have students observe and record what they hear. Then have the student holding the spoon place the ends of the string so they are touching the inside of the outer part of their ears, while the partner again taps the spoon. Then have team members switch roles. Ask students how the sound reached their ears. (It traveled through the string.)

4. Instruct one student to knock on a table. Have students record what they hear. Then have one team member place an ear on the table while the partner again knocks on the table. Discuss how the sound reached their ear. (It traveled through the wood.) Provide time for students to switch roles. Make sure students write their observations on their record sheets.

5. Challenge students to compare how sounds travel through air and solids. (Students may have noticed that sound travels faster through solids than through gases. It also sounds louder.)

Follow-Up

Ask students if they have ever heard a sound while swimming under water. Invite them to compare how sounds travel through solids, liquids, and gases.

Sound and Solids

Procedure and Observations

1. Tie a string to a spoon. Hold both ends of the string. Have your partner tap on the spoon. What do you hear?

   _____

   _____

   _____

2. Now hold the ends of the string to your ears. Have your partner tap on the spoon. What do you hear?

   _____

   _____

   _____

3. Have your partner knock on a table. What do you hear?

   _____

   _____

4. Now place your ear on the table. Have your partner knock on the table again. What do you hear?

   _____

   _____

Conclusion

5. What can you conclude about how sounds travel through gases (air) and solids?

   _____

   _____

   _____

Investigation 4

Changing Volume

Materials

- student record sheet on page 21, reproduced for each student
- combs
- craft sticks
- waxed paper

Steps to Follow

1. Clap your hands softly and then loudly. Ask students to identify the difference between the two sounds. Explain that **volume** is the loudness or softness of a sound. Invite students to identify some familiar soft and loud sounds. List them on the board under the headings "loud" and "soft."

2. Ask students what they think is different about creating a loud sound or a soft sound. Have them write their ideas on their record sheets.

3. Divide students into pairs. Distribute materials to each team.

4. Have students explore making loud and soft sounds. Tell them to tap loudly and then softly on a desk with a craft stick. Instruct them to record their observations on their record sheets.

5. Instruct students to hold the stick over the edge of the table as they did in Investigation 1. Challenge them to create a loud sound and a soft sound. Have them observe the motion of the stick during each sound and record their observations. (Bending the stick with more force creates a louder sound.)

6. Have students blow on a comb covered with waxed paper as they did in Investigation 1. Challenge them to create loud and soft sounds. Ask them what differences they notice in how it felt. Discuss how they created the different sounds. (Blowing harder created a louder sound.) Have students record their observations.

7. Discuss with students what they had to do in order to create sounds with different volumes. Ask them to compare the vibrations that made soft and loud sounds. (They had to create greater vibrations to produce a louder sound. They did this by inputting more energy.)

8. Have students look again at the list of sounds on the board. Discuss what is different about how these sounds are made. (More energy is needed to create the louder sounds.)

Follow-Up

Have students research oscilloscopes, devices that show the patterns of sound waves. Explain that the height (amplitude) of a sound wave determines the volume of the sound it produces. The higher the wave, the louder the sound.

Name ______________________________

Changing Volume

Concept 2 Energy
Investigation 4

Prediction

1. What do you think determines how loud or soft a sound is?

____

____

Procedure and Observations

2. Tap softly on a desk with the stick. Now tap loudly. On the chart below, record what you did to make each sound and what you observed.
3. Hold the stick over the edge of the table as you did in Investigation 1. Snap it to make a soft sound. Now snap it to make a loud sound. Record your observations.
4. Place waxed paper over a comb as you did in Investigation 1. Blow on the comb to make loud and soft sounds. Record your observations.

Action	Soft Sound		Loud Sound	
	What you did to make it	What you observed	What you did to make it	What you observed
Tapping stick				
Snapping stick				
Blowing on comb				

Conclusion

5. Based on your observations, what can you infer about the connection between energy, vibrations, and volume?

____

____

____

Investigation 5

Changing Pitch

Materials

- student record sheet on page 23, reproduced for each student
- craft sticks
- long pencils
- masking tape
- large plastic container with lid
- small plastic container with lid
- 30-cm rubber bands, thick, cut
- 30-cm rubber bands, thin, cut

Steps to Follow

1. Tap on a small plastic container. Then tap on a large plastic container. Ask students to describe the difference in the sounds. Write the word **pitch** on the board. Explain that pitch is how high or low a sound is.

2. Have students brainstorm a list of sounds with low pitch and high pitch. Write them in two lists on the board. Challenge students to predict how sounds of different pitch are created. Have them write their ideas on their record sheets.

3. Divide students into pairs. Have students wrap one end of a thick, cut rubber band twice around a craft stick and tape it in place. Have them wrap the other end of the rubber band twice around a pencil near the eraser and tape it in place.

4. Instruct a pair of students to hold the stick and the pencil (eraser-end down) upright on a desk so that the rubber band is parallel to the desk. Have them pluck the rubber band to create a sound. Show them how to roll the rubber band onto the pencil to shorten the distance between the stick and the pencil. Have them roll it up a small amount, pluck the rubber band, and compare the sounds. Tell them to repeat this several times until the rubber band is very short. Ask how their setup changed. Instruct students to record their observations on their record sheets.

5. Have students unroll the rubber band. This time, have one student gradually roll the rubber band onto the stick, while keeping the distance between the stick and the pencil constant. (The tension on the rubber band will increase.) After rolling it slightly, have them pluck the rubber band. Ask how their setup changed. Have students record their observations.

6. Provide students with a thinner, cut rubber band. Have them attach it to another stick and pencil in the same manner. Have two teams work together to set up their rubber bands next to each other so that they have the same length and tension. Have students pluck the two rubber bands, compare the sounds, and record their observations. Discuss what is different about the two setups.

7. Have students identify three factors that affect pitch. Encourage them to describe how length, tension, and thickness affect vibration rate and pitch. (Pitch increases with tension and decreases with length and thickness.)

Name _____

Changing Pitch

Prediction

1. How do you think sounds of different pitch are created?

_____

_____

Procedure and Observations

2. Wrap one end of a rubber band twice around a stick and tape it. Wrap the other end around a pencil and tape it.

3. Hold the stick and pencil so that the rubber band is parallel to your desk. Pluck the rubber band. Roll the rubber band onto the pencil to shorten its length. Pluck it again and compare the vibrations and sounds. Record your observations below.

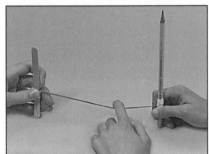

4. Unroll the rubber band. Then roll it back onto the stick but keep its length the same. (You'll increase the tension.) Pluck the rubber band. Record your observations.

5. Set up a thinner rubber band in the same manner. Set up the two rubber bands so they have the same length and tension. Pluck them. Record your observations.

Change in Rubber Band	Effect on Vibration Rate	Effect on Pitch
shortened length		
increased tension		
decreased thickness		

Conclusion

6. What relationship do you see between the rate of vibration and pitch?

_____

_____

Electrical energy flows through electric circuits.

Prepare in Advance

Investigation 1: This activity works best on a cool, dry day.

Teacher Information

All matter is made up of **atoms.** Atoms consist of a central nucleus surrounded by electrons. The nucleus contains protons, which are positively charged, and neutrons, which have no charge. The electrons, which orbit around the nucleus, are negatively charged. They remain at a certain distance from the nucleus and are held in orbit by the attraction between the negative and positive charges.

The arrangement of electrons around the nucleus in an atom determines whether or not that substance is a good **conductor.** Substances with atoms whose electrons are free to move about make good conductors of electric charge. Most metals are good conductors. Most plastics are poor conductors.

Neutral objects become negatively charged when their atoms gain extra electrons. They become positively charged when their atoms lose electrons. The buildup of charges on an object is known as **static electricity.** Eventually, the charges will leave the object in a momentary flow of electrons, known as **static discharge.**

An **electric current** is a continuous flow of electric charge. An **electric circuit** provides a pathway through which electric charge can flow. To keep current flowing in a circuit, energy is needed. A battery can provide an energy source.

An electrolyte is a liquid that conducts electricity. A wet cell is a battery that uses an electrolyte to conduct electrons and produce a flow of electric current. A dry cell is a battery that uses solid or paste-like chemicals to conduct current. When electrons flow from the negative end of a battery through various conductors to the positive end of the battery, a closed circuit is created.

In a series circuit, the circuit components (the battery, wires, and bulb) form a single path along which the electrons can move. As long as all connections are made in a circuit, electrons can continue to flow, and the circuit is called a **closed circuit.** Any break in the circuit causes the electrons to stop flowing. This is known as an **open circuit.**

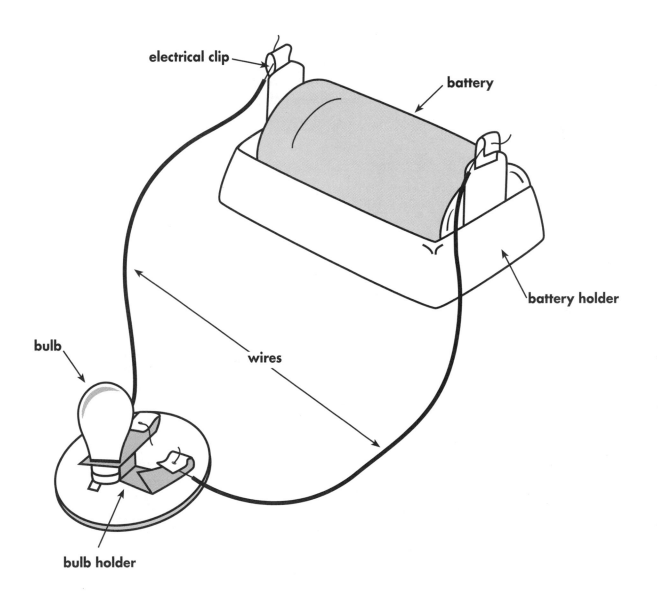

An Electric Circuit

electrical clip

battery

battery holder

bulb

wires

bulb holder

Concept
3
Energy

Investigation 1

Static Charges

Materials

See advance preparation on page 24.

- student record sheet on page 27, reproduced for each student
- 12" (30.5 cm) pieces of string
- round balloons
- wool fabric scraps

Steps to Follow

1. Ask students if they have ever gotten a shock when opening a door or pulling a sweater over their head. Invite them to describe these and similar experiences.

2. Divide students into pairs. Give each team two balloons and a piece of wool fabric. Instruct them to blow up the balloons, tie them, and attach a string to each. Tell students to rub one balloon with the fabric and then bring the fabric near the balloon. Ask them to observe what happens and write their observations on their record sheets.

3. Tell students to rub both balloons with the wool. Then have them hold the strings so that the two balloons hang close to each other. Ask them to observe and record what happens. Invite students to speculate as to why the balloons reacted differently in the two situations.

4. Remind students that all objects are made up of atoms. Explain that atoms are made up of smaller particles that have either positive or negative charges. When charges from one object interact with charges from another object, they produce electrical forces. These forces can be attractive or repellant. Encourage students to describe the movement of the balloons using the terms *attract, repel,* and *electric force.*

5. Explain that when they rubbed the balloon with wool, negatively charged particles, called electrons, were transferred from the fabric to the balloon. (Ask students which object was then positively charged and which was negatively charged.) Explain that the charge that resides on an object is known as **static electricity** because it is not moving.

6. Go on to explain that the fabric and balloon were attracted to each other because the balloon was relatively negatively charged and the fabric was relatively positively charged. Because the second balloon they rubbed also carried a negative charge, it repelled the first balloon.

Follow-Up

Have students touch the charged balloons to the floor and then observe their interaction again. (The balloons will not be attracted or repelled by one another.) Explain the concept of **static discharge.** (The built-up charges moved from the balloons to the floor. The balloons were then no longer charged.)

Static Charges

Procedure and Observations

1. Blow up the balloons. Tie them and attach a string to each.

2. Rub one balloon with the wool fabric. Bring the fabric close to the balloon. What happens? Record your observations.

   _____

   _____

3. Rub both balloons with the fabric. Hold the strings so the two balloons hang close to each other. What happens? Record your observations.

   _____

   _____

   _____

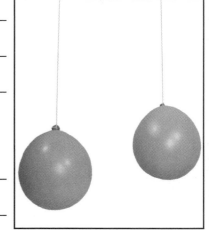

4. Again, rub the two balloons with the fabric. Hold them far apart. Slowly bring them closer to each other. What do you observe?

   _____

   _____

Conclusions

5. How can you explain the behavior of the balloon and the fabric?

   _____

   _____

   _____

6. How can you explain the behavior of the two rubbed balloons?

   _____

   _____

Concept 3 — Energy

Investigation 2
A Closed Circuit

Materials

- student record sheet on page 29, reproduced for each student
- battery holders with clips
- bulb holders
- D-cell batteries
- flashlight and batteries
- #14 flashlight bulbs
- insulated copper wires with stripped ends

Steps to Follow

1. Test the flashlight bulbs to make sure they work.

2. Hold up a flashlight. Turn it on and off. Invite students to explain what parts of the flashlight allow it to give off light. Take the flashlight apart. Hold up the battery and bulb. Ask students what their functions are. Explain that the battery is a source of electrical energy.

3. Touch the bulb to the battery. Make sure students notice that the bulb does not light up. Ask students what is needed to light the bulb.

4. Divide students into small groups. Give each group a battery, a wire, and a bulb. Challenge them to put the three together in as many ways as they can to light up the bulb. Have them draw their results on their record sheets.

5. Tell students that they have constructed an **electric circuit.** Explain that a simple electric circuit consists of a source of electrical energy (battery), an energy receiver (bulb), and a path that connects the two through which an electric current can move (wire). Ask students to identify the parts of their electric circuits.

6. Encourage students to make the bulb go on and off and record what they did.

7. Give each group a battery holder with two electrical clips, a bulb holder, and another piece of wire. Show students how to insert the ends of the wire into the clips on the battery holder and on the bulb holder. Challenge students to create an electric circuit.

Follow-Up

Encourage students to explain why the bulb is sometimes lit and sometimes not. Tell them that when there is an uninterrupted path for the current to follow, they have a **closed circuit.** Challenge them to define an **open circuit.** (The path for the current has been interrupted, so no energy can be transferred.)

Name ______________________________

Concept
3
Energy
Investigation 2

A Closed Circuit

Procedure and Observations

1. Use the wire, bulb, and battery to make the bulb light up. How many ways can you find that work? Draw each one.

2. How can you make the bulb go on and off? Record what you did.

____

____

3. Use a battery holder, a bulb holder, and a second wire. Make an electric circuit. Draw your circuit.

Conclusions

4. What is needed to make a bulb light up?

____

____

5. What is the difference between a closed circuit and an open circuit?

____

____

____

Investigation 3

Conductors

Materials

- student record sheet on page 31, reproduced for each student
- battery holders with clips
- bulb holders
- D-cell batteries
- #14 flashlight bulbs
- index cards
- insulated copper wires with stripped ends
- paper fasteners
- variety of metal and nonmetal objects

Steps to Follow

1. Test the flashlight bulbs to make sure they work.

2. Review with students what an electric current is (the flow of electrons through a closed electric circuit).

3. Ask students what materials they have seen an electric current flow through (wires, electrical clips, and bulb holder clips). Explain that a material through which an electric current can flow is called a **conductor.** Point out that all the materials they used in Investigation 2 were metals. Ask students if they think metals are the only materials that can conduct electricity. Ask them to predict what kinds of materials make the best conductors. Have them write their predictions on their record sheets.

4. Divide students into small groups. Have students build the same circuit they built in Investigation 2. Instruct them to add a conductor tester. (Poke two paper fasteners about 2" (5 cm) apart through an index card. One wire should go from the battery holder to one paper fastener. Another wire should go from the bulb holder to the other paper fastener. A third wire should connect the bulb holder to the battery holder. See diagram on the record sheet.) Explain that they now have an instrument to test materials to see if they are conductors.

5. Provide students with a variety of metal and nonmetal objects such as keys, coins, buttons, nails, pencils, strips of paper, rubber stoppers, string, plastic utensils, and aluminum foil. Tell them to test each object by placing it across the two paper fasteners. Ask students to explain how they will know if a tested material is a conductor or not. Have them record their answers.

6. Have students test the objects and record their observations. Discuss what kinds of materials make good conductors and what kinds do not.

Follow-Up

Ask students to explain why the wires they use are coated with plastic. Discuss the importance of both insulators (materials that do not conduct electric energy well) and conductors.

Conductors

Prediction

1. What kinds of materials do you think make good conductors?

_____

_____

Procedure and Observations

2. Set up the circuit as shown.

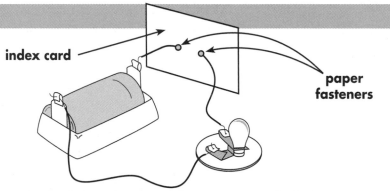

index card

paper fasteners

3. Test various objects to see which are good conductors. Record your test results below.

Material	Ability to Conduct Electric Current	Material	Ability to Conduct Electric Current

Conclusions

4. How can you tell if an object conducts electric current?

_____

_____

5. What kinds of materials make good conductors?

_____

_____

Concept **3** Energy

Investigation 4

A Wet Cell

Materials

- student record sheet on page 33, reproduced for each student

- alligator clips, with attached wire

- 250-mL beakers

- 1" x 4" (2.5 x 10 cm) copper strips or ¼"-thick (0.64 cm) copper wire

- galvanometers

- salt

- spoons

- water

- 1" x 4" (2.5 x 10 cm) zinc or aluminum strips or large galvanized zinc nails

Steps to Follow

1. Write the word **galvanometer** on the board. Show the students a galvanometer. Explain that it is a device used to test the amount of electric current in a circuit.

2. Ask students if they think they can create an electric current. Challenge them to tell how they would do it. Have them write their ideas on their record sheets.

3. Divide students into small groups. Have each group fill a beaker with about 7 oz. (200 mL) of warm tap water. Tell them to add several spoonfuls of salt to the water and stir until most of it dissolves.

4. Show students how to use an alligator clip to attach a copper strip to the inside rim of the beaker. Make sure that most of the strip is below the surface of the water. Have students do the same with the zinc strip on the other side of the beaker. Tell them to be sure that the two metal strips do not touch each other.

5. Instruct students to clip the free end of the alligator clip attached to the copper strip to the positive site on the galvanometer. Have them clip the other free alligator clip to the negative site on the galvanometer. Tell them to record their observations.

6. Challenge students to explain what is happening. Tell them that free electrons from the atoms in the copper strip are breaking away and moving across the water. Explain that these electrons are then picked up by the zinc, and the result is a flow of electric current.

Follow-Up

Have the students repeat the investigation using plain rather than salted water. (No electric current will flow.) Challenge them to explain the role of the salt water in this circuit. (It acts as a conductor between the zinc and copper.)

Have students try other liquid conductors. Provide them with lemons, lemon juice, apples, and potatoes. Discuss which liquid makes the best conductor and why.

A Wet Cell

Prediction

1. What do you think is needed to create an electric current?

   _____

   _____

Procedure and Observations

2. Fill a beaker with about 200 mL of warm tap water. Add salt and stir.

3. Use an alligator clip to attach the copper strip to the inside rim of the beaker so that most of the strip is in the water.

4. Clip the zinc strip to the other side of the beaker.

5. Clip the free end of the alligator clip attached to the copper strip to the positive site on the galvanometer. Clip the other free alligator clip to the negative site on the galvanometer.

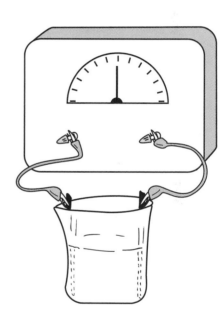

6. What happens when both wires are connected to the galvanometer?

   _____

   _____

   _____

Conclusions

7. What function do the two metal strips serve?

   _____

   _____

8. What function does the salt water serve?

   _____

   _____

Magnetism is related to electricity.

Prepare in Advance

Caution: Do not work with magnets around TVs, computers, computer discs, or cassette tapes, as these objects may be damaged by the magnets.

Teacher Information

A **magnet** has a north pole and a south pole. As with electric charges—in which like charges repel and opposite charges attract—like poles in a magnet repel and opposite poles attract. The **magnetic field** is the area around a magnet in which its force affects objects. The magnetic field is strongest at the poles and decreases in strength as you move farther from the poles. The Earth has its own magnetic field, which exerts a magnetic force that can be detected by compass needles.

Some materials, including iron, cobalt, and nickel, are naturally magnetic. These materials are attracted to magnets and can also be made into magnets. For example, you can make a temporary magnet by rubbing an iron nail with a magnet many times in the same direction.

Scientists have developed a model to explain how magnetism works. The electrons in atoms are constantly spinning. This spinning creates a tiny magnetic field around the electron. In most materials, the magnetic fields of individual atoms cancel each other out. In some materials, however, the atoms group together in areas called domains. The atoms within domains are aligned with one another so that they form a tiny magnet, complete with north and south poles. An object such as an iron nail has many magnetic domains. When all the magnetic domains line up in the same direction, the object acts like a magnet. This is why an iron nail can be made into a temporary magnet simply by stroking it with a permanent magnet. (The stroking in one direction aligns the domains.)

A solenoid is a coil of wire that carries an electric current. The current creates a magnetic field around the wire, with one end of the coil becoming the north pole of the magnet and the other end becoming the south pole. By changing the direction of the current, the poles can be reversed. A solenoid with a magnetic material inside it, such as iron, forms an **electromagnet.**

Conversely, an electric current can be produced by moving a coil of wire through a magnetic field. This is known as **electromagnetic induction.** A generator makes use of electromagnetic induction to produce electricity.

Magnetism

a) The arrows show the magnetic field around a magnet.

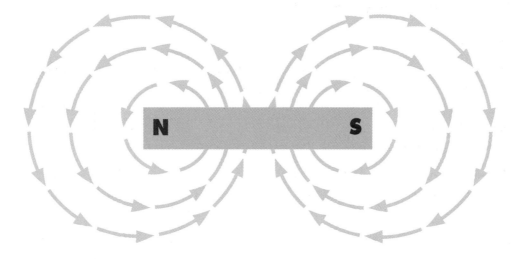

b) When an iron nail is stroked with a permanent magnet, all its magnetic domains line up, and it becomes a temporary magnet.

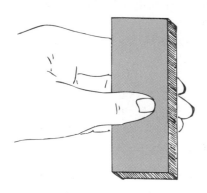

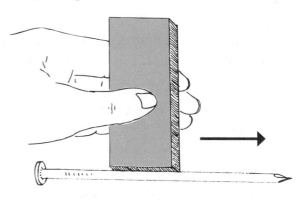

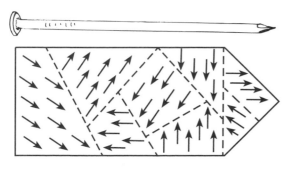

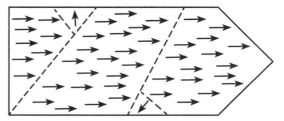

Concept **4** Energy

Investigation 1

Magnets and Energy

Materials

- student record sheet on page 37, reproduced for each student
- bar magnets
- various magnetic and nonmagnetic objects

Steps to Follow

1. Invite students to share their experiences with magnets.

2. Display various objects such as a pencil, paper clip, paper fastener, a piece of aluminum foil, an iron nail, and an eraser. Invite students to predict what will happen when a magnet is brought near each of these objects. Have them write their predictions on their record sheets.

3. Divide students into small groups. Provide students with bar magnets and various objects. Have them test each object by bringing the magnet close to it. Ask students to identify what kinds of materials are attracted to magnets. Have them record their observations.

4. Explain that the ability they observed of a magnet to exert a pull or a push on an object is called **magnetism.** Ask students whether they think magnetism is a form of energy. Challenge them to explain their answers. Point out that the pulling force of a magnet is evidence of energy—energy contained within the atoms that make up matter.

Follow-Up

Provide students with various kinds of magnets, including horseshoe magnets. Ask them to predict whether these magnets will have any differences in their ability to pick up objects. Have them test their predictions.

Encourage students to magnetize an object. Provide them with iron nails and have them stroke a nail in one direction repeatedly with one end of a bar magnet. Then have them test the ability of the nail to pick up paper clips or other small objects. Have them tap the nail on the table and try it again. Point out that some kinds of materials are permanent magnets, others can be made into temporary magnets, and some can never be magnets.

Magnets and Energy

Prediction

1. How do you think the objects your teacher has provided will respond to a magnet? Make a prediction about each object.

_____

_____

_____

Procedure and Observations

2. Bring the magnet close to each object. Note how each object reacts to the magnet.

Object	Response to Magnet	Object	Response to Magnet

Conclusions

3. What kinds of objects are attracted to a magnet?

_____

_____

4. What kinds of objects are not attracted to a magnet?

_____

_____

Investigation 2

Magnetic Fields

Materials

- student record sheet on page 39, reproduced for each student
- bar magnets
- box lids
- iron filings
- loose staples

Steps to Follow

1. Display a bar magnet. Ask students if they think the pulling force of the magnet is the same all over. Ask them to predict what differences might be observed and then write their predictions on their record sheets.

2. Have students work in pairs. Instruct them to spread a handful of staples on a desk. Tell them to place the bar magnet in the center of the staples. Have them pick up the magnet and notice where on the magnet staples are attached, and the quantity. Ask students to record their observations. Lead students to conclude that the magnetic force is strongest at the magnet's poles.

3. Have students place one bar magnet on the table. Have one student hold a second magnet and bring the north pole of this magnet near the north pole of the magnet on the table. Then have them bring the north pole of one magnet near the south pole of the second magnet. Have them bring the south poles of the two magnets together. Challenge them to explain what they have observed.

4. Instruct students to put a bar magnet on the table and place a box lid over it, with the edges up. Show them how to sprinkle iron filings lightly on the lid. Tell them to hold one corner of the lid so it cannot slip and then tap gently at the side. Ask students to describe what happens. Have them draw what they see on their record sheets and label the poles of the magnet.

   Caution: Instruct students to handle the iron filings with care and to keep them away from the magnets.

5. Challenge students to explain what they observed. Write the term **magnetic field** on the board. Explain that a magnetic field is the area around a magnet that exerts a push or pull on objects. Ask students where they think the magnetic field was strongest. Have them compare this to their observations of the magnet's ability to pick up staples.

Follow-Up

Ask students whether they think the two poles of a magnet have equal strength. Encourage them to carry out an experiment to find out.

Challenge students to predict the shape of the magnetic field around two bar magnets placed side by side. Have them test their predictions using iron filings.

Name _____

Magnetic Fields

Prediction

1. How do you think the pulling force of a magnet is distributed throughout the magnet?

_____

_____

Procedure and Observations

2. Spread a handful of staples on a desk. Set the magnet in the center of the staples. Then pick it up. What do you observe?

_____

_____

3. Put a bar magnet on the table. Place a box lid over it, with the edges up. Sprinkle iron filings lightly on the lid.

4. Hold one corner of the lid so it cannot slip. Gently tap the side. What happens? On the back of this sheet, draw the pattern made by the iron filings. Label the poles of the magnet.

Conclusions

5. Where is the magnetic field of a bar magnet strongest?

_____

6. How does the area with the strongest magnetic field compare with your observations of the magnet's ability to pick up staples? Explain.

_____

_____

_____

Concept 4 Energy

Investigation 3

Electric Current and Magnetism

Materials

- student record sheet on page 41, reproduced for each student
- 1.5-volt (AA) batteries
- bulb holders
- compasses
- #14 flashlight bulbs
- short and long pieces of insulated copper wires with stripped ends
- sheets of paper
- bar magnets

Steps to Follow

1. Test the flashlight bulbs to make sure they work.

2. Divide students into small groups. Give each group a compass. Invite them to explore how the compass needle moves. Ask what they think the needle is made of.

3. Distribute the rest of the materials. Instruct students to put a bar magnet on a sheet of paper and trace around it to make sure the magnet is kept in the same position. Then have them place the compass anywhere on the paper. Have them draw an arrow to show the direction in which the needle points. Tell them to repeat this at five or six different spots around the paper.

4. Lead students to conclude that the bar magnet is interacting with the needle of the compass, which is magnetized. Depending on where the compass is placed, its needle points in a different direction. (The needle should always point toward the south pole of the magnet.)

5. Ask students to predict how an electric current placed near a compass will affect the compass needle. Have students write their predictions on their record sheets.

6. Have students insert a bulb into the holder. Have them use a battery and two wires (one short, one long) to create a closed circuit. Make sure their bulbs light up. Tell students that the lit bulb shows that electricity is flowing through the circuit.

7. Tell students to disconnect the longer wire from the battery. Have them hold this wire over the compass so that it is parallel to the compass needle. **Caution them not to let the wire touch the needle.** Instruct one group member to briefly touch the end of the wire to the battery while the others watch the needle.

   Caution: Tell students to just briefly touch the wire to the battery, as it will quickly get very warm. Short circuits happen easily and dead batteries may result.

8. Invite students to explain what they observed. Ask them why they think the wire with a current running through it was able to affect the needle. Explain that current running through a wire produces a magnetic field around the wire. That magnetic field affected the compass needle.

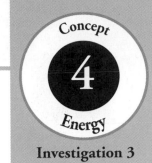

Electric Current and Magnetism

Prediction

1. How do you think an electric current placed near a compass will affect a compass needle?

_____

_____

Procedure and Observations

2. Put a bar magnet on a sheet of paper and trace around it. Place the compass anywhere on the paper. Draw an arrow to show the direction in which the needle points.

3. Repeat Step 2 from five or six different spots around the paper.

4. Put the bulb into the holder. Set up a closed circuit like the one shown.

5. Disconnect the longer wire from the battery. Hold the wire over the compass so that it is parallel to the compass needle. *Do not let the wire touch the needle.*

6. Briefly touch the end of the wire to the battery. How does the electric current in the wire affect the compass needle?

_____

_____

Conclusions

7. How is a compass needle affected by a magnetic field?

_____

_____

8. Why does an electric current affect a compass needle?

_____

_____

_____

Concept 4 Energy

Investigation 4

An Electromagnet

Materials

- student record sheet on page 43, reproduced for each student
- battery holders with clips
- 1.5-volt (AA) batteries
- iron nails
- long insulated wires with stripped ends
- paper clips
- tape

Steps to Follow

1. Ask a volunteer to review how an electric current affects a compass needle.

2. Ask students when a wire acts like a magnet (when an electric current is moving through it).

3. Explain to students that they will use a wire with an electric current moving through it to make a powerful magnet called an **electromagnet.**

4. Divide students into small groups. Distribute materials to each group.

5. Instruct students to wrap tape around four iron nails. Then have them wrap 10 coils of wire around the nails. Tell them to make sure that each coil of wire touches the next one.

6. Have students attach the ends of the wire to the battery holder clips. Instruct them to bring the nails close to a pile of paper clips. Have them note what happens.

   Caution: Tell students not to leave the circuit connected for too long. Short circuits happen easily and dead batteries may result.

7. Invite students to suggest how they can make their electromagnet stronger. Steer them toward suggesting that the number of coils of wire might affect the strength of the magnet. Have students write their predictions on their record sheets.

8. Have students test their predictions. Tell them to wrap the wire around the nails 10 more times and test how many paper clips their electromagnet will pick up.

9. Have students repeat Step 8 two more times.

10. Discuss how the number of coils of wire affects the strength of the electromagnet. Challenge students to explain their answers. Reinforce the idea that electric current can create magnetic force.

Follow-Up

Ask students what they think would happen if they increased the amount of current in the wire. Discuss how this could be accomplished. Have students test it out.

An Electromagnet

Prediction

1. How do you think you could make the electromagnet stronger?

_____

_____

Procedure and Observations

2. Tape four iron nails together. Wrap 10 coils of wire around the nails. Make sure each coil touches the next one.

3. Attach the ends of the wire to the battery holder clips.

4. Bring the nails close to a pile of paper clips. What happens? Record the number of paper clips affected on the chart below.

5. Wrap the wire around the nails 10 more times. How many paper clips do the nails pick up now? Record the number.

6. Repeat Step 5 two more times, each time adding 10 more coils.

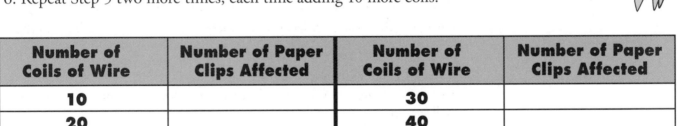

Number of Coils of Wire	Number of Paper Clips Affected	Number of Coils of Wire	Number of Paper Clips Affected
10		30	
20		40	

Conclusions

7. How does the number of coils of wire around the nails affect the strength of the electromagnet? Why do you think this happens?

_____

_____

8. What other factors do you think affect the strength of the electromagnet?

_____

_____

Investigation 5

Electromagnetic Induction

Materials

- student record sheet on page 45, reproduced for each student

- insulated wire with stripped ends

- galvanometers

- strong and weak bar magnets

Steps to Follow

1. Invite a volunteer to review how a magnetic field can be created from an electric current flowing through a wire. Then ask students if they think electricity can be created using a magnet. Encourage them to suggest how this could be done.

2. Write the word **galvanometer** on the board. Remind students that a galvanometer is an instrument used to measure the amount of electric current in a circuit. Tell students that in this activity they will use a galvanometer to test for the presence of a small electric current.

3. Divide students into small groups. Distribute materials to groups.

4. Have students loop the wire into at least 7 coils. (The coils must be large enough to accommodate a bar magnet that will be inserted later.) Then demonstrate how to connect the uninsulated ends of the wire to the terminals on the galvanometer. Have students record the galvanometer reading on their record sheets.

5. Tell students to move the end of a bar magnet halfway into the coils of wire. Have them again record the galvanometer reading.

6. Have students observe and record the galvanometer reading with each of the following variables: moving the magnet in and out of the coils quickly, moving the magnet slowly, moving the magnet farther into the wire coils, using a stronger bar magnet, and increasing the number of coils.

7. Discuss with students what they have demonstrated. Explain to them that the process of producing an electric current by moving a coil of wire through a magnetic field is called **electromagnetic induction.**

8. Challenge students to explain all their observations. Ask them to evaluate which variables affected the process the most and which affected it the least. Encourage them to explain why some variables were more effective than others.

Follow-Up

Summarize this concept by asking students how electricity and magnetism are related. Make sure they realize that a magnetic field can be used to induce an electric current and an electric current can produce a magnetic field.

Name _____

Electromagnetic Induction

Procedure and Observations

1. Loop the wire into at least 7 coils. Connect the ends of the wire to the galvanometer. Record the reading on the chart below.

2. Move the end of a bar magnet halfway into the coils of wire. Record the reading.

3. Make each of the changes shown below. Record your observations.

Variable Tested	Galvanometer Reading
Moving magnet in and out of the wire coils	
Moving the magnet quickly	
Moving the magnet slowly	
Moving the magnet farther into the wire coils	
Using a stronger bar magnet	
Increasing the number of coils in the wire	

Conclusions

4. How can you explain what happened when you moved the magnet in and out of the coiled wire?

_____

_____

5. Which variables affected the process of electromagnetic induction the most?

_____

_____

_____

Heat energy can be transferred.

Prepare in Advance

Investigation 1: Prepare a large container of cold water and another of hot water.

Investigation 3: Melt the bottom of each candle and stand it upright in an aluminum pie pan.

Teacher Information

When people speak of heat, most of the time they are really talking about thermal energy. Scientists define heat as the transfer of thermal energy. Since students at this grade level are more familiar with the term *heat*, we will use that term in these investigations.

Heat is caused by the motion of molecules that make up matter. All molecules are constantly in motion, but the speed of their motion varies. The faster the molecules are moving, the hotter the substance is.

Heat flows from warmer objects to cooler objects. Coolness is simply the absence of heat. Heat can be transferred by three principal means: **conduction, convection,** and **radiation.**

When molecules bump into each other, heat energy is transferred from the faster-moving molecules to the slower-moving molecules. As they gain heat energy, the slower-moving molecules begin to speed up. They in turn bump into other slower-moving molecules and speed them up. Heat energy is transferred from one molecule to another until all the molecules in a substance are moving at the same speed. This process is known as conduction. A metal spoon sitting in a cup of hot tea is warmed by conduction.

Heat convection occurs in liquids and gases. When a liquid or gas gains heat energy, the molecules move faster and begin to spread out. Because the heated gas or liquid is less dense than the unheated substance that surrounds it, it begins to rise. The resulting currents of liquid or gas are called **convection currents.**

Radiation is the transfer of energy through electromagnetic waves. (See Concept 1.) The sun heats the Earth through electromagnetic radiation.

Temperature is a measure of how hot or cold something is. It is not a measure of the amount of heat in an object, because this measurement depends upon the mass of the substance. However, temperature is a measure of how fast the molecules of a substance are moving, and is thus related to heat.

Three Methods of Heat Transfer

Conduction

Convection

Radiation

Concept **5** Energy

Investigation 1

Does Heat Move?

Materials

See advance preparation on page 46.

- student record sheet on page 49, reproduced for each student
- beakers
- cold water
- hot water
- shallow pans
- thermometers

Steps to Follow

1. Ask students whether they feel warmer when they are sitting still or when they are exerting a lot of energy. Explain that **heat** is a form of energy. Tell students that heat energy results from moving molecules. Explain that the warmer a substance is, the faster its molecules are moving.

2. Divide students into small groups. Show them the classroom supply of cold and hot water. Have students fill two beakers with the same amount of hot water. Then have them fill one pan half full with cold water and the second pan half full with hot water.

3. Have students measure and record the temperature of the water in each beaker. (They should be the same.)

4. Instruct students to set one beaker in the pan of hot water and the second beaker in the pan of cold water. Ask students to predict what will happen to the temperature of the water in the beakers. Provide time for them to write their predictions on their record sheets.

5. Instruct students to measure and record the water temperature in each beaker every 5 minutes for at least 20 minutes.

6. Discuss what happens to the water temperature in each beaker over time. Encourage students to explain where the movement of molecules is faster and where it is slower.

Follow-Up

Have students graph their data. Suggest that they label time on the horizontal axis and temperature on the vertical axis.

Have students fill two beakers with cold water. Have them fill one pan half full with cold water and a second pan half full with hot water. Instruct them to measure and record the temperature in each beaker every 5 minutes for at least 20 minutes. Have students compare these results to their previous data.

Name _____

Concept
5
Energy

Investigation 1

Does Heat Move?

Prediction

1. What do you think will happen to the temperatures in the beakers of water over time?

_____

_____

Procedure and Observations

2. Fill two beakers with hot water. Fill one pan half full with cold water and another pan half full with hot water.

3. Measure and record the temperature of the water in each beaker at the start.

4. Set one beaker in the pan of hot water. Place the second beaker in the pan of cold water.

5. Every 5 minutes, measure the water temperature in each beaker. Record your data on the chart below.

Beaker of Cold Water		Beaker of Hot Water	
Time	Temperature (°C)	Time	Temperature (°C)

Conclusions

6. What happens to the temperature of hot water placed in a bath of cold water over time?

_____

7. What happens to the temperature of hot water placed in a bath of the same temperature water over time?

_____

8. What can you conclude about the behavior of heat energy?

_____

Investigation 2

Conduction

Materials

- student record sheet on page 51, reproduced for each student
- beakers
- hot water
- metal spoons, long- and short-handled
- plastic spoons
- wax or shortening

Steps to Follow

1. Review with students what they learned about heat from the previous investigation. Explain that heat can flow in several different ways. In this investigation, they will investigate heat **conduction.**

2. Divide students into small groups. Distribute materials to each group.

3. Instruct groups to fill a beaker with hot water. Tell them to place a small ball of wax or dab of shortening on the tip of the handle of each of the three spoons.

4. Ask students to predict what will happen when they place the spoons in the hot water. Have them write their predictions on their record sheets.

5. Have students place all three spoons in the hot water so that the handles extend beyond the beaker and are not directly over the hot water. Caution them not to let the spoons touch each other. Allow several minutes for student observations. Tell students to touch the spoon handles and notice how they feel. Make sure they record their observations.

6. After several minutes of observations, encourage students to describe their observations. (Students should find that the wax or shortening has melted on the metal spoons, but not on the plastic spoon.)

7. Offer the following explanation of student results: The molecules of water are moving very fast because the water is hot. These fast-moving molecules bump into the molecules of the spoon. This causes the molecules in the spoon to get more heat energy and move faster. The molecules at the bottom of the spoon hit those above it and transfer the heat energy, and this process continues all the way up the spoon.

8. Explain that a material through which heat energy can be easily transferred is called a **conductor.** Ask which is a better conductor, plastic or metal? Challenge students to explain their answers in terms of the molecules that make up each material.

Follow-Up

Encourage students to repeat the activity using various other materials, such as a wooden spoon or stick, a straw, and spoons or sticks made from different metals. Have them time how long it takes for the wax to melt. Then have them rate the quality of each material as a conductor.

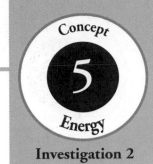

Concept

5

Energy

Investigation 2

Conduction

Prediction

1. What do you think will happen to the wax on the end of the three spoon handles when the spoons are put in hot water?

_____

_____

Procedure and Observations

2. Fill a beaker with hot water. Place a small ball of wax or shortening on the tip of the handle of each spoon.

3. Put all three spoons into the hot water. Wait several minutes. What happens?

_____

_____

4. Touch the spoon handles. How do they feel?

_____

Conclusions

5. Which spoon is made of a material that is a better conductor?

_____

_____

6. How is heat transferred from the water to the wax?

_____

_____

7. How can you explain any differences in your observations between the long-handled and the short-handled metal spoons?

_____

_____

Investigation 3

Convection

Materials

See advance preparation on page 46.

- student record sheet on page 53, reproduced for each student
- aluminum foil
- aluminum pie pans, holding votive candles
- heavy books
- matches
- rulers or doweling
- safety goggles
- scissors
- thread
- transparent tape
- white paper

Steps to Follow

1. Ask a volunteer to review how heat moves by conduction. Explain that in this activity they will investigate another kind of movement of heat.

2. Divide students into small groups. Have each group draw a spiral on a sheet of white paper. Tell them to place the paper on top of a piece of aluminum foil and trace over the lines with a pen. Make sure they press hard as they trace. Then have them cut along the lines they made on the foil. Check to make sure they have created spirals.

3. Tell students to tape one end of a piece of thread to the center of the spiral. Have them tie the other end of the thread to the ruler or doweling.

4. Have students extend the stick with the spiral over the edge of a desk. Direct them to place a book over the stick to hold it in place. Check to make sure that the spirals hang down to a height of about 12" (30.5 cm) from the floor.

5. Instruct students to place their candles (in pans) under the spiral. Ask them to predict what will happen when they light the candle. Tell them to record their predictions.

   Caution: Warn students to use extreme care when working with matches and candles.

6. Have students put on their safety goggles. Allow students to light their candles. Have them observe what happens to the spiral and record their observations. Then tell them to blow out the candles.

7. Ask students to describe their observations. Challenge them to explain why the spiral moved. Be sure they understand that the heat from the flame warmed the air molecules above it. This heated mass of air molecules began to expand and rise. Explain that such movement of heated molecules is known as **convection.** Tell students that the moving air molecules then pushed against the spiral and caused it to spin.

8. Save the candles in the pans for use in Investigation 4.

Follow-Up

Ask students to observe convection in liquids. Have them put a colored hard candy or a drop of food coloring in the bottom of a beaker of water and then gently heat the water. Challenge students to explain what they see. Encourage them to discuss how the heat is transferred throughout the water.

Convection

Prediction

1. What do you think will happen when you light the candle under the spiral?

_____

_____

_____

_____

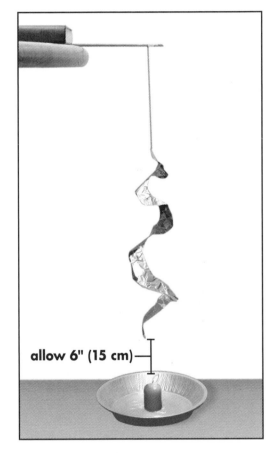

allow 6" (15 cm)

Procedure and Observations

2. Tape one end of a piece of thread to the center of the aluminum foil spiral you made. Tie the other end of the thread to a ruler.

3. Place the ruler with the spiral over the edge of a desk. Put a book over the ruler to hold it in place.

4. Put on your safety goggles. Place the candle under the spiral, allow 6" (15 cm) of space between the spiral and the top of the candle. If it is too close, trim your spiral. Light the candle.

5. What happened when you lit the candle under the spiral?

_____

_____

Conclusion

6. How can you explain the movement of the spiral?

_____

_____

_____

_____

 Energy • EMC 879

Investigation 4

Radiation

Materials

- student record sheet on page 55, reproduced for each student
- aluminum pie pans, holding candles
- marshmallows
- matches
- metal forks
- safety goggles
- thermometers

Steps to Follow

1. Ask students to name the two kinds of heat transfer they have already investigated (conduction and convection). Tell them they will now investigate a third kind of heat transfer.

2. Divide students into small groups. Distribute the materials to each group.

3. Direct students to hold a thermometer about 4" (10 cm) above the unlit candle. Have them measure and record the air temperature.

4. Have students stick a marshmallow onto the end of a metal fork.

5. Provide each student with a pair of safety goggles. Demonstrate how to light the candle.

   Caution: Warn students to use extreme care when working with matches and candles.

6. Have students hold the marshmallow about 4" (10 cm) above the candle. Caution them to keep the marshmallow well above the top of the flame. Have them observe what happens and record their observations.

7. Have students hold the thermometer about 4" (10 cm) above the candle. Have them measure and record the air temperature. Caution them not to leave the thermometer there very long. Then tell students to blow out the candles.

8. Invite volunteers to describe what happened to the marshmallow. Ask them to describe what their results showed about the transfer of heat energy.

9. Explain to students that what they have observed is the transfer of heat through **radiation.** Point out that the flame did not touch the marshmallow, and yet it got cooked anyway. Radiation does not require direct contact between materials.

10. Explain that when students sit in the sun on a sunny day, they can feel the sun's heat. Explain that this is another example of radiation.

11. Save the pans with candles for use in Concept 6, Investigation 3.

Follow-Up

Invite students to summarize the similarities and differences between conduction, convection, and radiation. Challenge them to make a list of places in their homes and in school where heat is transferred in each of these ways.

Radiation

Procedure and Observations

1. Hold a thermometer about 10 cm above the candle. Measure and record the air temperature.

   Temperature above unlit candle: _____

2. Stick a marshmallow onto the end of a metal fork.

3. Put on safety goggles. Light the candle.

4. Hold the marshmallow about 10 cm above the candle. Make sure to keep the marshmallow well above the top of the flame. What happens?

   _____

   _____

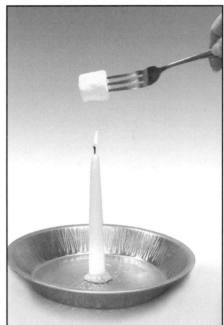

5. Once again, hold the thermometer about 10 cm above the candle. Keep it well above the top of the flame for a very short time. Measure and record the air temperature.

   Temperature above burning candle: _____

6. Blow out the candle.

Conclusions

7. What do your results show about the transfer of heat energy?

   _____

   _____

8. How is radiation different from convection and conduction? How is it the same?

   _____

   _____

   _____

   _____

Energy can change from one form to another.

Prepare in Advance

Investigation 1: Use ethyl alcohol in place of isopropyl alcohol if it is available, as it provides more distinctive results.

Investigation 5: Remove the ends from cardboard cylinders such as oatmeal boxes or tin cans. If you use tin cans, tape the ends with masking tape to cover up any sharp edges.

Teacher Information

Energy is the ability to do work. There are five basic categories of energy. One of these is **mechanical energy.** Any object in motion has mechanical energy. For example, a ball flying through the air, water in a waterfall, and a galloping horse all have mechanical energy. Sound energy is a type of mechanical energy.

Heat energy results from the motion of atoms and molecules. The faster these particles move, the more heat energy a substance contains. The addition or subtraction of heat energy leads to a change in temperature.

Chemical energy exists in the bonds that hold atoms together. When these bonds are broken, chemical energy is released. The chemical energy in food is released when the food is digested and the chemical bonds are broken.

Electric charges in motion contain energy. Electric energy can be carried in an electric current through a wire.

Nuclear energy is contained in the nucleus of an atom. This energy can be released when the nucleus is split, or when certain nuclei collide at high speeds and fuse. A nuclear fusion reaction produces the sun's energy.

The **law of conservation of energy** states that energy can be transformed from one form to another, but it cannot be created or destroyed. All forms of energy can be converted to other forms. Green plants convert the energy in sunlight to chemical energy through the process of **photosynthesis.** The chemical energy in wood is converted to heat and light energy when it is burned. Chemical energy stored in a battery is converted to electrical energy when the battery is placed in a circuit.

Burning wood is an **exothermic** reaction. It gives off heat energy. Photosynthesis is an **endothermic** reaction. It requires heat in order to occur.

Energy Conversion

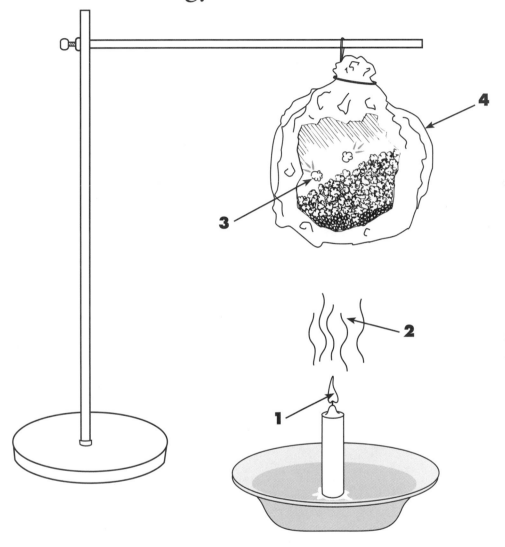

1) The stored chemical energy in the candle is converted to light and heat energy when lit.

2) Heat energy radiates to the aluminum foil bag.

3) The heat energy causes a chemical reaction in the popcorn. Some of the chemical energy in the popcorn is converted to the mechanical energy of movement.

4) The mechanical energy of the moving popcorn kernels causes them to strike the bag. This causes vibrations in the aluminum foil, which are converted to sound waves, another form of mechanical energy.

Investigation 1

Photosynthesis

Materials

See advance preparation on page 56.

- student record sheet on page 59, reproduced for each student
- aluminum foil
- droppers or pipettes
- ethyl or isopropyl alcohol
- geranium plant
- hot plate
- iodine
- large beaker
- paper clips
- paper towels
- petri dishes
- safety goggles
- tweezers

Steps to Follow

1. Brainstorm with the class a list of energy forms. Write the list on the board.

2. Display a large healthy geranium plant. Explain that this and other green plants are energy converters. Invite students to speculate as to what kind of energy conversion green plants can do. Have them write their predictions on their record sheets.

3. Divide students into small groups. Have each group use paper clips to attach a piece of foil to a different leaf of the plant. Make sure the foil covers only part of the leaf, and that no light reaches that part. Ask why the leaves are partially covered (to compare parts that receive light with parts that do not).

4. Place the plant in bright light for at least four days.

5. Wearing goggles, gently heat a beaker half full of ethyl alcohol on a hot plate. Have each group remove a leaf from the plant and take off the foil. Put the leaves in the beaker. Cook them for several minutes until limp. Use tweezers to take the leaves out of the beaker and place them on a paper towel to dry.

6. Give each group a leaf on a petri dish. Tell students that iodine is an indicator of starch, a food. It turns black if starch is present.

7. Wearing safety goggles, have students use a dropper to completely cover the leaf with iodine. Tell them to wait several minutes and then record their observations. If no color change has occurred, have them observe the leaf again after about an hour.

   Caution: Iodine is poisonous and can stain clothing. Advise students to use extreme caution when handling it.

8. Have students compare the results on the parts of the leaf that were covered and uncovered. (The covered parts will contain little if any food and should not stain black.) Challenge them to explain any differences. Invite them to explain how plants are energy converters. Make sure they understand that plants convert the energy in sunlight to chemical energy through the process of **photosynthesis.**

Photosynthesis

Prediction

1. What kind of energy conversion do you think occurs in green plants?

_____

_____

Procedure and Observations

2. Attach a piece of foil to part of a leaf.

3. After four days, remove a leaf and its foil and have your teacher heat it in ethyl alcohol.

4. Cover the leaf with iodine.

5. Describe how the leaf reacted to iodine.

The covered part:

_____

_____

The uncovered part:

_____

_____

Conclusions

6. How can you explain any differences in the reaction of the two parts of the leaf to iodine?

_____

_____

7. What conclusions about the conversion of light energy by plants can you draw from your observations?

_____

_____

         Energy • EMC 879

Concept 6 Energy

Investigation 2

Chemical Energy Is a Gas

Materials

- student record sheet on page 61, reproduced for each student
- balloons
- double-acting baking powder
- measuring spoons
- newspapers
- paper towels
- small containers, such as film canisters
- vinegar

Steps to Follow

1. Ask students what kind of energy conversion they observed in the first investigation (light energy to chemical energy). Explain that in this investigation they will observe what happens in a reaction when chemical energy is released.

2. Divide the students into small groups. Provide newspapers for students to spread at their workstations to catch spills.

3. Give each group two small containers and have them fill each about half full with vinegar. Then direct them to pour the vinegar from each container into a balloon.

4. Instruct one group member to hold the balloons while another dries out the containers with a paper towel. Have students put ¼ tsp. baking powder into one container and ½ tsp. baking powder into the second container.

5. Show students how to stretch the balloon over the top of a container without letting the vinegar fall out of the balloon. Have them put a balloon over each container. Ask students to predict what they think will happen when the vinegar mixes with the baking powder in the two containers. Have them record their predictions on their record sheets.

6. Tell students to dump the vinegar from each balloon into its container. Have them observe what happens and record their observations.

7. Ask students to identify the type of energy conversion that occurred. Tell students that the chemical reaction that took place between the vinegar and the baking powder produced carbon dioxide gas, which filled up the balloon, causing it to expand (mechanical energy).

Follow-Up

Invite students to brainstorm other situations in which chemical energy is converted to mechanical energy, the energy of motion.

Chemical Energy Is a Gas

Prediction

1. What do you think will happen when the vinegar mixes with the baking powder in the containers?

_____

_____

Procedure and Observations

2. Fill each container half full with vinegar.
   Pour the vinegar from each container into a balloon.

3. Dry out the containers with a paper towel.
   Put ¼ tsp. baking powder into one container.
   Put ½ tsp. baking powder into the second container.

4. Stretch a balloon over the top of each container.
   Do not let the vinegar enter the container.

5. Dump the vinegar from each balloon
   into its container.

6. What happened when you dumped the
   vinegar into the baking powder? Draw a
   picture of each container and balloon
   to the right.

Conclusions

7. What difference did you note between the balloons on each container? How can you explain your observations?

_____

_____

8. What kind of energy conversion occurred?

_____

_____

Investigation 3

Chemical Energy Pops

Materials

- student record sheet on page 63, reproduced for each student
- overhead transparency of *Energy Conversion* on page 57
- 6" (15 cm) aluminum foil squares
- aluminum pie pans, holding candles (from Concept 5)
- matches
- paper clips
- popcorn kernels
- ring stands
- safety goggles

Steps to Follow

1. Display a candle. Ask students what kind of energy is stored in the candle wax (chemical energy). Explain that in this activity that energy will be converted.

2. Give each pair of students a square of aluminum foil and 10 to 15 kernels of popcorn. Instruct students to place the popcorn on the foil and twist the corners of the foil together to make a bag. Tell students to be sure to leave a lot of air space in the bag.

3. Direct students to straighten the middle loop in a paper clip and hook one end through the foil bag. Provide each team with a ring stand and have them hang the paper clip and bag over the cross bar.

4. Make sure all students are wearing safety goggles. Have students place the unlit candle under the bag. The bag should hang about 2 to 3" (5 to 8 cm) above the candlewick.

5. Ask students to predict what kind of energy changes will take place when they light the candle. Have them write their predictions on their record sheets.

   Caution: Warn students to use extreme care with open flames.

6. Tell students to light the candle and then listen to what happens. When they can't hear any more sounds, have them blow out the candle and let the bag cool. Then have them open the bag and make visual observations.

7. Using the *Energy Conversion* transparency, discuss the kinds of energy conversions that took place in this activity. Make sure students are aware that all of the following energy changes are occurring: chemical energy of candle to heat and light energy of flame, to mechanical energy of popping corn, to sound energy when corn hits the aluminum.

Follow-Up

Invite students to identify other situations in which chemical energy is converted to heat or light, and in which mechanical energy is converted to sound.

Chemical Energy Pops

Prediction

1. What kind of energy change or changes do you think will occur when you light a candle under a bag of popcorn?

_____

_____

Procedure and Observations

2. Place the popcorn on the foil. Twist the corners of the foil together to form a bag. Make the bag much bigger than the popcorn.

3. Straighten a paper clip in the middle. Hook one end through the bag. Use the hook to hang the bag from the ring stand. *Caution: Use extreme care with open flames.*

4. Put on safety goggles. Place a candle under the bag. Leave 5 to 8 cm of space between the bag and the candle.

5. Light the candle. What happened when you lit the candle?

_____

_____

6. Blow out the candle. Let the bag cool.

7. Open the bag. What happened to the popcorn inside the bag?

_____

_____

Conclusion

8. How was energy converted in this experiment?

_____

_____

_____

Investigation 4

Heat Energy

Materials

- student record sheet on page 65, reproduced for each student
- seltzer tablets
- small beakers
- thermometers
- water

Steps to Follow

1. Divide students into small groups. Give each group a beaker. Have them fill it about one-quarter full with lukewarm water. Have students feel the bottom of the beaker and describe what they observe. Ask them what the temperature tells them (the average speed of the molecules of water). Ask what kind of energy is present in the water (heat energy).

2. Tell students to measure and record the temperature of the water.

3. Show students a seltzer tablet. Invite them to predict what will happen to these tablets when they are placed in water. Have them write their predictions on their record sheets.

4. Direct students to break up two seltzer tablets and drop them into the water. Instruct them to again feel the beaker and measure the water temperature. Remind them to record their observations.

5. Invite students to explain what kind of energy conversion happened. Explain that a chemical reaction occurred that used energy from the water. As a result, the water temperature was lower after the reaction. Chemical reactions that absorb heat energy are called **endothermic** reactions.

6. Encourage students to discuss what happened in terms of the speed of the water molecules. (The molecules slowed down as they lost heat energy.)

Follow-Up

Tell students that some chemical reactions, known as **exothermic** reactions, release heat energy. Encourage students to do research about other chemical reactions and to identify them as endothermic or exothermic.

Name ____________________

Concept 6 Energy

Investigation 4

Heat Energy

Prediction

1. What do you think will happen when seltzer tablets are placed in warm water?

____

____

Procedure and Observations

2. Fill a beaker about one-quarter full with lukewarm water. Touch the bottom of the beaker. Record what you feel on the chart below.
3. Measure and record the water temperature on the chart below.
4. Break up two seltzer tablets. Drop them into the water.
5. Touch the bottom of the beaker. Record your observations.
6. Measure and record the water temperature.

	Observations	**Temperature (°C)**
Before adding seltzer tablets		
After adding seltzer tablets		

Conclusions

7. What does a change in the temperature of a beaker of water tell you about the amount of heat energy in it?

____

____

8. Describe the conversion of energy in this experiment.

____

____

____

Investigation 5

Sound Energy

Materials

See advance preparation on page 56.

- student record sheet on page 67, reproduced for each student
- cardboard or metal cylinders
- round balloons
- rubber bands
- salt
- scissors

Steps to Follow

1. Remind students that in a previous activity they observed heat energy being converted to mechanical energy and then to sound energy (popcorn). Tell students that in this investigation they will investigate one way that sound energy is converted to another form of energy.

2. Divide students into pairs. Distribute materials to each team.

3. Instruct students to cut the top off the balloon and stretch it over one end of the cylinder. Emphasize that the balloon should be tightly stretched. Tell students to hold the balloon in place by putting a rubber band around the balloon and the cylinder.

4. Have students sprinkle salt on the stretched balloon. Instruct one team member to hold the cylinder, keeping it level.

5. Ask students what they think will happen when a student bends down and shouts into the open end of the cylinder. Have them write their predictions on their record sheets.

6. Instruct students to test their predictions. Make sure the student holding the cylinder observes what happens. (The salt will dance on the balloon.) Have students record their observations.

7. Have teammates reverse positions and repeat the procedure.

8. Challenge students to identify the energy conversions that they observed. (Sound energy was converted to mechanical energy.)

Follow-Up

Have students again sprinkle salt on the balloon. Have one student hold the cylinder while a second student holds a metal pan about 6 to 8" (15 to 20 cm) above it and taps the pan with a wooden spoon or stick. Tell students to observe what happens. (The salt will dance on the balloon again.) Invite them to explain their observations.

Sound Energy

Prediction

1. What do you think will happen when you bend down and shout into the open end of the cylinder?

_____

_____

Procedure and Observations

2. Cut off the top of the balloon. Stretch the balloon over one end of a cylinder. Make sure it is tightly stretched. Put a rubber band around the balloon and cylinder to hold the balloon in place.

3. Sprinkle salt on the balloon. Have your partner hold the cylinder level.

4. Shout into the open end of the cylinder. Have your partner observe what happens to the salt grains. Record your observations.

_____

_____

_____

5. Change places with your partner. Repeat Steps 3 and 4.

6. Describe what you see happening to the salt grains.

_____

_____

_____

Conclusion

7. Explain what happened to the salt grains in terms of energy and energy conversion.

_____

_____

_____

Investigation 6

Electric Energy

Materials

- student record sheet on page 69, reproduced for each student
- battery holders with clips
- D-cell batteries
- disc magnets
- electrical clips
- emery cloths
- enamel-coated copper wires, 4.3' (1.3 m) long
- insulated copper wires with the ends stripped, 10" (25 cm) long
- plastic wrap
- slabs of clay

Steps to Follow

1. Display a battery. Ask students to recall what kind of energy is produced by a battery (electrical). Tell the class that in this activity they will observe the conversion of electrical energy to another form of energy.

2. Divide students into small groups. Give each group a disc magnet and a length of enamel-coated copper wire. Instruct students to use the wire to make a coil around the magnet about 2" (5 cm) from the end of the wire. Tell them to wrap the wire around the magnet 14 more times.

3. Tell students to remove the magnet from the coils. Show them how to wrap each end of the wire around the coil to hold it in place.

4. Have students lay the wire coil on the desk and straighten out the ends. Provide teams with a piece of emery cloth to remove the enamel from the top half of the wire at each end. Tell them not to remove the enamel from the bottom half of the wire.

5. Tell students to construct the circuit shown in the figure on their record sheets by inserting two electrical clips into the slab of clay, placing a disc magnet between the clips, and balancing the coil between the clips. Attach one end of each copper wire to one of the electrical clips in the clay. Tell them not to connect the wires to the battery yet.

6. Have students check to make sure the coil is balanced. Have them flick the coil with their fingers to make sure it can spin.

7. Ask students what they think will happen when they close the circuit. Have them write their predictions on their record sheets.

8. Direct students to connect both wires to the clips on the battery holder. Have them flick the coil again to start it spinning. Tell them to observe and record what happens, and then to disconnect the wires.

9. Challenge them to explain what kind of energy conversion occurred (electric to mechanical). Ask what they think they have built (a motor).

10. Explain that the battery created an electric current in the coil, which then produced a magnetic field around the coil. Point out that this temporary magnetic field interacted with the magnetic field of the permanent magnet below it, causing the coil to spin.

Electric Energy

Prediction

1. What do you think will happen when you close your circuit?

_____

_____

Procedure and Observations

2. Follow your teacher's instructions to make a coil.

3. Place the wire coil on a desk or table. Straighten out the ends. Use a piece of emery cloth to rub off the enamel from the top half of the wire at each end. ***Leave the enamel on the bottom half of the wire.***

4. Put together the circuit as directed by your teacher. Do not connect the wires to the battery holder.

5. Balance the coil in the electrical clips. Tap it with your finger to make sure it can spin.

6. Connect both wires to the clips on the battery holder. Tap the coil again to start it spinning. What happens? Record your observations.

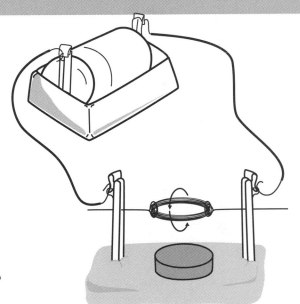

_____

_____

7. Disconnect the wires.

Conclusion

8. What kind of energy conversion did you observe?

_____

_____

_____

Energy can change matter.

Prepare in Advance

Investigation 1: Separate the eggs and discard the yolks. Keep one white and one yolk separate for display purposes.

Investigation 2: Prepare ice cubes and crushed ice. Fill a bucket with ice water. Place the butyl stearate in a pail of warm water so that it will thaw. For each team, fill a reclosable bag about one-quarter full with the liquid butyl stearate. Reseal the bags and place them in a bucket of ice water. Fill reclosable bags about one-quarter full with crushed ice. Place them in the bucket as well. Allow time for the butyl stearate to solidify. Prepare a classroom supply of warm water.

Teacher Information

Matter changes form all the time. Some of these are **physical changes,** changes that do not produce a new substance. Physical changes include breaking something, grinding it, or moving it from one size container to another.

Physical changes also include changes of state. There are three principal states of matter: solid, liquid, and gas. When a solid changes to a liquid, or melts, a physical change occurs. The same is true when a liquid changes to a gas, or vaporizes. Both of these processes can happen in reverse, because the matter has not changed to a new substance. For example, when heat energy is added, ice can change to water and then to water vapor, but the water vapor can be changed back into water or ice by taking away heat energy. These changes can happen many times, but the substance remains H_2O. Its physical properties, including density, melting point, and boiling point, do not change. Neither do its chemical properties, which describe the ways it interacts with other substances.

A **chemical change** differs from a physical change in that it causes a substance to change into one or more other substances. When a chemical change occurs, the new substances have different physical and chemical properties than the original substance. The original substance cannot be restored simply by reversing the process. For example, when crystallized sugar is heated, it first melts and then begins to burn. As it burns, the sugar compound, which contains carbon, hydrogen, and oxygen atoms, breaks down and forms water molecules and carbon molecules. When the carbon and water are cooled, they do not turn back into sugar molecules. The physical and chemical properties of the water and carbon molecules are different from those of the sugar molecules.

Butyl stearate is a substance that melts at room temperature (70°F). This makes it a nice example of how different substances have different melting points.

Physical and Chemical Changes

Physical Change

Physical Change

Chemical Change

Chemical Change

Concept **7** Energy

Investigation 1

Physical Changes

Materials

See advance preparation on page 70.

- student record sheet on page 73, reproduced for each student

- eggbeaters or wire whisks

- egg whites

- shallow bowls

Steps to Follow

1. Show students an egg that has been separated into the white and the yolk. Invite students to predict what will happen if the egg white is beaten vigorously with an eggbeater. Ask them if they think the egg white will still be an egg white or if it will become a different substance. Have students write their predictions on their record sheets.

2. Divide students into small groups. Give each group an egg white, an eggbeater, and a bowl. Instruct them to observe the appearance of the egg white. Make sure they note the color, texture, volume, and any other important features. Have them record their observations on their charts.

3. Tell students to beat the egg vigorously. Have them again observe and record the appearance of the egg white. Make sure they identify all the changes they observe.

4. Ask students if they think the substance they see now is the same as what they started with. Challenge them to explain their reasons.

5. Tell students to let the egg white stand. Ask what they think will happen. Have them make observations every 5 minutes. Remind them to record their observations and to note all changes that occur each time. (The egg white should start to return to its original state.)

6. Encourage students to explain what happened to the egg white. Ask them if the substance that is left in the container at the end is the same as the substance they started with. Invite them to explain their reasons. Ask them how they could further test their ideas.

7. Explain to students that they have observed a physical change. Tell them that a physical change is any change in a substance that does not produce a new substance.

Follow-Up

Invite students to again beat the egg white to see if it returns to the same state again. Discuss why this happens. Ask students if there is a limit to the number of times they can beat the egg white and still have it remain an egg white. (There is no limit.)

Challenge students to name other physical changes they may have observed. Encourage them to explain how they know that a physical change has occurred.

Physical Changes

Prediction

1. What do you think will happen to an egg white if it is beaten vigorously?

_____

Procedure and Observations

2. Observe the appearance of the egg white. Note its color, volume, and texture. Record your observations on the chart below.

3. Beat the egg white vigorously. What happens? Record your observations below.

4. Let the egg white stand. Make and record your observations of its appearance every 5 minutes.

Time	Appearance of Egg White
Before beating	
After beating	
5 minutes later	
10 minutes later	
15 minutes later	
20 minutes later	
25 minutes later	

Conclusions

5. After beating the egg white vigorously, is the substance you see the same as what you started with? Explain why or why not.

_____

_____

6. Is the substance that is left in the container after 25 minutes the same as the substance you started with? Explain why or why not.

_____

_____

7. Did a physical change occur? How can you tell? Record your answer on the back of this sheet.

Investigation 2

Solids and Liquids

Materials

See advance preparation on page 70.

- student record sheet on page 75, reproduced for each student

- beakers

- classroom supply of warm water

- clock or watches

- paper towels

- reclosable bags containing butyl stearate

- reclosable bags containing crushed ice

- thermometers

Steps to Follow

1. Ask students to identify the three states of matter (solids, liquids, and gases). Invite them to give examples of materials that represent each state of matter.

2. Give each pair of students a bag of crushed ice and a bag of butyl stearate. Invite students to describe what is in each bag. Write the term **butyl stearate** on the board.

3. Ask students what they think will happen to each substance if heat energy is added to it. Have them write their predictions on their record sheets.

4. Have each team fill two beakers about half full with warm water.

5. Have students open each bag and insert a thermometer into the solid. Make sure they push the thermometer deep into the solid substance. Instruct them to tightly reseal the bags and place one bag in each beaker.

6. Have students measure the temperature on each thermometer and observe the appearance of the material in each bag. Tell them to record their observations on the chart on the line labeled "Start."

7. Instruct students to repeat Step 6 every few minutes. Make sure they also record the time. Have them continue making observations until the solid has melted and the temperature has started to rise again. Tell them to replace the water in the beakers if it cools off.

8. Invite students to explain what they have observed. Point out that as long as a substance is changing state, from solid to liquid in this case, the temperature remains constant. Tell students that this plateau temperature represents the melting point of the substance.

9. Ask students to explain why they put the bags into warm water. Make sure they understand that the water added heat energy to the solids.

10. Ask students if they think the change that has occurred is a physical change. Encourage them to describe how they could test their ideas.

Follow-Up

Challenge students to explain why temperature plateaus while a substance is melting. (The heat energy is being used to cause the physical change. Once the substance has melted, its temperature begins to rise again.)

Concept

7

Energy

Investigation 2

Solids and Liquids

Prediction

1. What do you think will happen to the solids when heat energy is added?

_____

_____

Procedure and Observations

2. Fill two beakers about half full with warm water.

3. Open each bag and push a thermometer into the solid. Reseal the bags.

4. Place a bag in each beaker.

5. Observe the temperature on each thermometer. Record the data on the chart below. Also observe and record the appearance of the material in each bag.

6. Continue to record the temperature and appearance of each solid every 3 or 4 minutes. Record the time of each observation.

Time	Crushed Ice		Butyl Stearate	
	Temperature (°C)	Appearance	Temperature (°C)	Appearance
Start				

Conclusion

7. What happened to the temperature and appearance of each substance? How can you explain this?

_____

_____

_____

Investigation 3

Liquids and Gases

Materials

- student record sheet on page 77, reproduced for each student
- beakers
- hot plates
- thermometers
- warm water

Steps to Follow

1. Review the physical change that students observed in the previous activity (melting). Ask what other kind of physical change involves a change of state.

2. Divide students into small groups. Have students fill a beaker with about 5 oz. (150 mL) of warm water. Tell them to put the thermometer in the water and measure and record the water temperature.

3. Have students place the beaker on an electric hot plate and turn on the heat.

   Caution: Warn students to be careful not to touch the hot plate or the beaker after they have been heated.

4. Ask students to predict what will happen to the water in the beaker. Have them write their predictions on their record sheets.

5. Tell students to measure and record the temperature every minute or two. Have them also record the appearance of the water.

6. Once the water begins to bubble, have students continue to make observations for 5 to 10 more minutes. Then have them turn off the hot plate.

7. Have students note and record the volume of water that remains in the beaker. Ask them what happened to the water that is gone. (It changed to water vapor and entered the air.)

8. Challenge students to compare this activity to the previous one. Discuss what was similar and what was different. Point out that again a physical change occurred, and that during the time the water was changing state, the temperature did not rise. Make sure students understand that this is because the heat energy was being used to fuel the process of evaporation.

9. Challenge students to explain how they could test to be sure that this was a physical change. (They would need to capture the escaping water vapor and cool it back into water.)

Follow-Up

Ask students how many times they could change water to gas and back again and not have it change into a different substance (as many as they'd like). Make sure they understand that when a physical change occurs, the substance remains the same.

Liquids and Gases

Prediction

1. What do you think will happen to the water when it is heated?

   _____

   _____

Procedure and Observations

2. Fill a beaker with 150 mL of warm water. Measure and record the temperature of the water on the chart below.

3. Place the beaker on the hot plate. Turn on the heat. ***Caution: Do not touch the hot plate or the beaker.***

4. Measure and record the water temperature every minute or two. Also record the appearance of the water.

5. When you are finished making observations, turn off the hot plate.

6. Observe and record the volume of water that remains in the beaker. _____

Time	Water Temperature (°C)	Appearance of Water

Conclusions

7. How was this activity the same as the previous one?

   _____

8. How was this activity different from the previous one?

   _____

Concept **7** Energy

Investigation 4

Chemical Changes

Materials

- student record sheet on page 79, reproduced for each student
- black construction paper
- hot plates
- magnifiers
- metal pans
- metal spoons
- safety goggles
- granulated sugar

Steps to Follow

1. Display a container of sugar. Tell students what it is. Have students sprinkle a few grains of sugar on a sheet of black construction paper. Tell them to observe it with a magnifier and to record what they see.

2. Ask students what they think will happen to the sugar when it is heated. Have them write their predictions on their record sheets.

3. Divide students into small groups. Distribute the materials.

4. Have students put a spoonful of sugar into a metal pan and place it on a cool hot plate. Instruct them to put on safety goggles.

   Caution: Warn students not to touch the hot plate or the pan.

5. Have students turn the heat to medium low. Have them continuously stir the sugar and observe any changes. Remind them to observe smells as well as visual changes. Tell students to record all their observations. When they think a gas is beginning to form, direct them to turn off the heat.

6. Invite students to describe the changes they observed in the sugar. Ask them if they think a physical change occurred. Challenge them to explain their answers. Remind them that when a physical change occurs, the substance remains the same. Point out that the sugar first melted, which was a physical change, but then other changes occurred.

7. Ask them if they think the white sugar will return if they cool the substance. Have them allow the pan to cool and observe the substance again. Tell them to record what they see.

8. Discuss the results. Explain to students that they have seen an example of a **chemical change.** Invite them to try to define chemical change. Make sure they understand that during a chemical change there is a reaction that causes a change from one substance to another.

Follow-Up

Encourage students to describe other chemical changes they have observed around them.

Name _____

Chemical Changes

Prediction

1. What do you think will happen to the sugar when it is heated?

_____

_____

Procedure and Observations

2. Sprinkle sugar on a sheet of black construction paper. Observe it with a magnifier. Describe the appearance of the sugar at the start.

_____

3. Put a spoonful of sugar into a metal pan. Place it on a cool hot plate.

4. Put on safety goggles. ***Caution: Do not touch the hot plate or the pan.***

5. Turn the heat to medium low. Stir the sugar constantly. Observe any changes. What changes occurred in the sugar when it was heated?

_____

6. When you think a gas is beginning to form, turn off the heat.

7. Allow the pan to cool. Observe the substance again. What did the substance look like after it cooled?

_____

Conclusions

8. Do you think a physical change occurred in the sugar? Why or why not?

_____

_____

9. What evidence did you see that a chemical change occurred?

_____

_____